COMMERCIAL PILOT KNOWLEDGE TEST GUIDE

U.S. DEPARTMENT OF TRANSPORTATION
FEDERAL AVIATION ADMINISTRATION
Flight Standards Service

Preface

FAA-G-8082-5, Commercial Pilot Knowledge Test Guide, provides information for obtaining authorization to take the commercial pilot knowledge tests. Appendix 1 provides lists of reference materials and subject matter knowledge codes, and a list of computer testing designees (CTD's).

Changes to the subject matter knowledge codes will be published in AC 60-25, Reference Materials and Subject Matter Knowledge Codes for Airman Knowledge Testing.

The current Flight Standards Service airman training and testing material, questions banks, and subject matter knowledge codes for all airman certificates and ratings can be obtained from the Regulatory Support Division, AFS-600, home page on the Internet.

The Regulatory Support Division's Internet address is: www.mmac.jccbi.gov/afs/afs600

FAA-G-8082-5 supersedes Advisory Circular (AC) 61-114, dated 1995, and can be purchased from the Superintendent of Documents, U.S. Government Printing Office (GPO), Washington, DC 20402-9325, or from U.S. Government Bookstores located in major cities throughout the United States. For an explanation of why the Commercial Pilot Knowledge Test Guide was taken out of the AC system, refer to AC 60-29, Renumbering of Airman Training and Testing Publications.

Comments regarding this guide should be sent to the Federal Aviation Administration, Airman Testing Standards Branch, AFS-630, Attn: Commercial Pilot Certification Area Manager, P.O. Box 25082, Oklahoma City, OK 73125.

Contents

APPENDIX 1

INTRODUCTION

What is required to become a skilled and competent commercial pilot? Although some individuals possess more knowledge and skills than others, no one is a natural-born pilot. Competent commercial pilots become so through study, training, and experience.

This knowledge test guide will answer most of your questions about taking a commercial pilot knowledge test. It will cover the following areas: knowledge test eligibility requirements; knowledge areas on the tests; descriptions of the tests; process for taking a knowledge test; use of test aids and materials; cheating or other unauthorized conduct; validity of Airman Test Reports; retesting procedures; and practical test eligibility requirements.

- This guide will help in preparing you to take one or all of the following knowledge teCommercial Pilot—Airplane
- Commercial Pilot—Rotorcraft–Helicopter
- Commercial Pilot—Rotorcraft–Gyroplane
- Commercial Pilot—Glider
- Commercial Pilot—Balloon–Hot Air
- Commercial Pilot—Balloon–Gas
- Commercial Pilot—Lighter-Than-Air–Airship
- Military Competency—Airplane
- Military Competency—Helicopter

This guide is not offered as an easy way to obtain the necessary information for passing the knowledge tests. Rather, the intent of this guide is to define and narrow the field of study to the required knowledge areas included in the tests.

Rote memorization of test questions may render good test scores, but a correlative understanding of the subject matter may be lacking. This correlative understanding of the entire aviation environment is what produces a safe and effective pilot.

KNOWLEDGE TEST ELIGIBILITY REQUIREMENTS

To be eligible to take the knowledge test, you must have:

1. Received an endorsement from an authorized instructor certifying that you have accomplished a ground-training or a home-study course required by this part for the certificate or rating sought and you are prepared for the knowledge test; and

2. Proper identification at the time of application that contains your—

   a. Photograph;

   b. Signature;

   c. Date of birth, which shows you meet or will meet the age requirements of this part for the certificate sought before the expiration date of the Airman Test Report; and

   d. Actual residential address, if different from your mailing address.

KNOWLEDGE AREAS ON THE TESTS

Commercial pilot tests are comprehensive because they test your knowledge in many subject areas. If you are pursuing a commercial pilot certificate, you should review the appropriate knowledge areas listed below, pertinent to the category sought. Airship applicants should review instrument procedures. Balloon and airship applicants should include a review of the fundamentals of instructing.

Aeronautical knowledge areas:

- Applicable Code of Federal Regulations (CFR's) that apply to commercial pilot privileges, limitations, and flight operations.
- Accident reporting requirements of the National Transportation Safety Board.
- Basic aerodynamics and the principles of flight.
- Meteorology to include recognition of critical weather situations, wind shear recognition and avoidance, and the use of aeronautical weather reports and forecasts.
- The safe and efficient operation of aircraft.
- Weight and balance computation.
- Use of performance charts.
- Significance and effects of exceeding aircraft performance limitations.
- Use of aeronautical charts and magnetic compass for pilotage and dead reckoning.
- Use of air navigation facilities.
- Aeronautical decision making and judgment.
- Principles and functions of aircraft systems.
- Maneuvers, procedures, and emergency operations appropriate to the aircraft.
- Night and high altitude operations.
- Procedures for operating within the National Airspace System.
- Procedures for flight and ground training for lighter-than-air ratings.

DESCRIPTIONS OF THE TESTS

All test questions are the objective, multiple-choice type. Each question can be answered by the selection of a single response. Each test question is independent of other questions; therefore, a correct response to one does not depend upon, or influence, the correct response to another. The minimum passing score is 70 percent.

Communication between individuals through the use of words is a complicated process. In addition to being an exercise in the application and use of aeronautical knowledge, a commercial pilot knowledge test is also an exercise in communication since it involves the use of written language. Since the tests involve written rather than spoken words, communication between the test writer and the person being tested may become a difficult matter if care is not exercised by both parties. Consequently, considerable effort is expended to write each question in a clear, precise manner. Make sure you carefully read the instructions given with each test, as well as the statements in each test item.

- The following tests each contain 100 questions, and you are allowed 3 hours to complete each tesCommercial Pilot—Airplane
- Commercial Pilot—Rotorcraft–Helicopter
- Commercial Pilot—Rotorcraft –Gyroplane
- Commercial Pilot—Glider
- Commercial Pilot—Balloon–Hot Air
- Commercial Pilot—Lighter-Than-Air–Airship

The following test contains 60 questions, and you are allowed 2.5 hours to complete the test.

- Commercial Pilot—Balloon–Gas

The following tests each contain 50 questions, and you are allowed 2 hours to complete each test.

- Military Competency—Airplane
- Military Competency—Helicopter

PROCESS FOR TAKING A KNOWLEDGE TEST

The Federal Aviation Administration (FAA) has available hundreds of computer testing centers worldwide. These testing centers offer the full range of airman knowledge tests including military competence, instrument foreign pilot, and pilot examiner screening tests. Refer to appendix 1 of this guide for a list of computer testing designees (CTD's).

The first step in taking a knowledge test is the registration process. You may either call the central 1-800 numbers (refer to appendix 1 for 1-800 numbers) or simply use the walk-in basis. If you choose to use the 1-800 number to register, you will need to select a testing center, schedule a test date, and make financial arrangements for test payment. You may register for tests several weeks in advance, and you may cancel your appointment according to the CTD's cancellation policy. If you do not follow the CTD's cancellation policies, you could be subject to a cancellation fee.

The next step in taking a knowledge test is providing proper identification. You should determine what knowledge test prerequisites are necessary before going to the computer testing center. Your instructor or local Flight Standards District Office (FSDO) can assist you with what documentation to take to the testing facility. Testing center personnel will not begin the test until your identification is verified. A limited number of tests do not require authorization. An endorsement from an authorized instructor is not required for military competency.

Acceptable forms of authorization:

- A certificate of graduation or a statement of accomplishment certifying the satisfactory completion of the ground school portion of a course from an FAA-certificated pilot school.

- A certificate of graduation or a statement of accomplishment certifying the satisfactory completion of the ground school portion of a course from an agency such as a high school, college, adult education program, U.S. Armed Force, ROTC Flight Training School, or Civil Air Patrol.

- A written statement or logbook endorsement from an authorized instructor certifying that you have accomplished a ground-training or home-study course required for the rating sought and you are prepared for the knowledge test.

- Failed Airman Test Report, passing Airman Test Report, or expired Airman Test Report (pass or fail), provided that you still have the original Airman Test Report in your possession.

Before you take the actual test, you will have the option to take a sample test. The actual test is time limited; however, you should have sufficient time to complete and review your test.

When taking a test, keep the following points in mind:

- Answer each question in accordance with the latest regulations and guidance publications.

- Read each question carefully before looking at the possible answers. You should clearly understand the problem before attempting to solve it.

- After formulating an answer, determine which test answer corresponds with your answer. The answer chosen should completely resolve the problem.

- From the answers given, it may appear that there is more than one possible answer; however, there is only one answer that is correct and complete. The other answers are either incomplete, erroneous, or represent common misconceptions.

- If a certain question is difficult for you, it is best to mark it for review and proceed to the next question. After you answer the less difficult questions, return to those which you marked for review and answer them. The review procedure will be explained to you prior to starting the test. Although the computer should alert you to unanswered questions, make sure every question has an answer recorded. This procedure will enable you to use the available time to the maximum advantage.

- When solving a calculation problem, select the answer closest to your solution. The problems have been checked manually and with various types of calculators. If you have solved it correctly, your answer will be closer to the correct answer than any of the other choices.

Upon completion of the knowledge test, you will receive your Airman Test Report, with the testing center's embossed seal, which reflects your score.

The Airman Test Report lists the subject matter knowledge codes for questions answered incorrectly. The total number of subject matter knowledge codes shown on the Airman Test Report is not necessarily an indication of the total number of questions answered incorrectly. Appendix 1 contains a list of subject matter knowledge codes that refer to the knowledge areas. Study these knowledge areas to improve your understanding of the subject matter.

Your instructor is required to provide instruction on each of the knowledge areas listed on your Airman Test Report and to complete an endorsement of this instruction. You must present the Airman Test Report to the examiner prior to taking the practical test. During the oral portion of the practical test, the examiner is required to evaluate the noted areas of deficiency.

Should you require a duplicate Airman Test Report due to loss or destruction of the original, send a signed request accompanied by a check or money order for $1 payable to the FAA. Your request should be sent to the Federal Aviation Administration, Airmen Certification Branch, AFS-760, P.O. Box 25082, Oklahoma City, OK 73125.

Use of Test Aids and Materials

Airman knowledge tests require applicants to analyze the relationship between variables needed to solve aviation problems, in addition to testing for accuracy of a mathematical calculation. The intent is that all applicants are tested on concepts rather than rote calculation ability. It is permissible to use certain calculating devices when taking airman knowledge tests, provided they are used within the following guidelines. The term "calculating devices" is interchangeable with such items as calculators, computers, or any similar devices designed for aviation-related activities.

1. Guidelines for use of test aids and materials. The applicant may use test aids and materials within the guidelines listed below, if actual test questions or answers are not revealed.

a. Applicants may use test aids, such as scales, straightedges, protractors, plotters, navigation computers, log sheets, and all models of aviation-oriented calculating devices that are directly related to the test. In addition, applicants may use any test materials provided with the test.

b. Manufacturer's permanently inscribed instructions on the front and back of such aids listed in 1(a), e.g., formulas, conversions, regulations, signals, weather data, holding pattern diagrams, frequencies, weight and balance formulas, and air traffic control procedures are permissible.

c. The test proctor may provide calculating devices to applicants and deny them use of their personal calculating devices if the applicant's device does not have a screen that indicates all memory has been erased. The test proctor must be able to determine the calculating device's erasure capability. The use of calculating devices incorporating permanent or continuous type memory circuits without erasure capability are prohibited.

d. The use of magnetic cards, magnetic tapes, modules, computer chips, or any other device upon which prewritten programs or information related to the test can be stored and retrieved are prohibited. Printouts of data will be surrendered at the completion of the test if the calculating device used incorporates this design feature.

e. The use of any booklet or manual containing instructions related to the use of the applicant's calculating device is not permitted.

f. Dictionaries are not allowed in the testing area.

g. The test proctor makes the final determination relating to test materials and personal possessions that the applicant may take into the testing area.

2. Guidelines for dyslexic applicant's use of test aids and materials. A dyslexic applicant may request approval from the local Flight Standards District Office (FSDO) to take an airman knowledge test using one of the three options listed in preferential order:

a. Option One. Use current testing facilities and procedures whenever possible.

b. Option Two. Applicants may use Franklin Speaking Wordmaster® to facilitate the testing process. The Wordmaster® is a self-contained electronic thesaurus that audibly pronounces typed in words and presents them on a display screen. It has a built-in headphone jack for private listening. The headphone feature will be used during testing to avoid disturbing others.

c. Option Three. Applicants who do not choose to use the first or second option may request a test proctor to assist in reading specific words or terms from the test questions and supplement material. In the interest of preventing compromise of the testing process, the test proctor should be someone who is non-aviation oriented. The test proctor will provide reading assistance only, with no explanation of words or terms. The Airman Testing Standards Branch, AFS-630, will assist in the selection of a test site and test proctor.

CHEATING OR OTHER UNAUTHORIZED CONDUCT

Computer testing centers must follow strict security procedures to avoid test compromise. These procedures are established by the FAA and are covered in FAA Order 8080.6, Conduct of Airman Knowledge Tests. The FAA has directed testing centers to terminate a test at any time a test proctor suspects a cheating incident has occurred. An FAA investigation will then be conducted. If the investigation determines that cheating or unauthorized conduct has occurred, then any airman certificate or rating that you hold may be revoked, and you will be prohibited for 1 year from applying for or taking any test for a certificate or rating under 14 CFR part 61.

VALIDITY OF AIRMAN TEST REPORTS

Airman Test Reports are valid within the 24-calendar month period preceding the month you complete the practical test. If the Airman Test Report expires before completion of the practical test, you must retake the knowledge test.

Retesting procedures

If you receive a grade lower than a 70 percent and wish to retest, you must present the following:

- failed Airman Test Report; and

- a written endorsement from an authorized instructor certifying that additional instruction has been given, and the instructor finds you competent to pass the test.

If you decide to retake the test in anticipation of a better score, you may retake the test after 30 days from the date your last test was taken. The FAA will not allow you to retake a passed test before the 30-day period has lapsed. Prior to retesting, you must give your current Airman Test Report to the test administrator. The last test taken will reflect the official score.

PRACTICAL TEST ELIGIBILITY REQUIREMENTS

There are prerequisites for a commercial pilot certificate that require you to have a combination of experience, knowledge, and skill. For specific information pertaining to certification, you should carefully review the appropriate sections of Title 14 of the Code of Federal Regulations (14 CFR) part 61.

Additionally, to be eligible for a commercial pilot certificate, you must:

1. Be at least 18 years of age.

2. Be able to read, speak, write, and understand the English language. If you are unable to meet one of these requirements due to medical reasons, then the Administrator may place such operating limitations on your pilot certificate as necessary for the safe operation of the aircraft.

3. Receive a logbook endorsement from an authorized instructor who:

a. Conducted the required ground training or reviewed the person's home study on the aeronautical knowledge areas listed in 14 CFR part 61, section 61.125(b) that apply to the aircraft category and class rating sought; and

b. Certified that the person is prepared for the required knowledge test that applies to the aircraft category and class rating sought.

4. Pass the required knowledge test on the aeronautical knowledge areas listed in 14 CFR part 61, section 61.125(b).

5. Receive the required training and a logbook endorsement from an authorized instructor who:

a. Conducted the training on the areas of operation listed in 14 CFR part 61, section 61.127(b) of this part that apply to the aircraft category and class rating sought; and

b. Certified that the person is prepared for the required practical test.

6. Meet the aeronautical experience requirements of Subpart F that apply to the aircraft category and class rating sought before applying for the practical test.

7. Hold at least a valid third-class medical certificate issued under 14 CFR part 67. No medical certificate is required for a glider or balloon rating.

SAMPLE TEST QUESTIONS AND ANSWERS

AIRPLANE

1. A pilot flying a single-engine airplane observes a multiengine airplane approaching on a collision course from the left. Which pilot should give way?

A—Each pilot should alter course to the right.

B—The pilot of the single-engine airplane should give way; the other airplane is to the left.

C—The pilot of the multiengine airplane should give way; the single-engine airplane is to its right.

Answer C—Subject Matter Knowledge Code: B08. 14 CFR part 91, section 91.113 states:

"Sec. 91.113 Right-of-way rules: Except water operations....

(d) Converging. When aircraft of the same category are converging at approximately the same altitude (except head-on, or nearly so), the aircraft to the other's right has the right-of-way. If the aircraft are of different categories—

(1) A balloon has the right-of-way over any other category of aircraft;

(2) A glider has the right-of-way over an airship, airplane, or rotorcraft; and

(3) An airship has the right-of-way over an airplane or rotorcraft...."

2. The ratio between the total load supported by the wing and the actual weight of an aircraft in flight is known as

A—load factor and directly affects stall speed.

B—aspect load and directly affects stall speed.

C—load factor and has no relation with stall speed.

Answer A—Subject Matter Knowledge Code: H303. AC 61-23C, Pilot's Handbook of Aeronautical Knowledge; Chapter 1: Loads and Load Factors, page 1-18.

A load factor is the ratio of the total airload acting on the airplane to the gross weight of the airplane. For example, a load factor of 3 means that the total load on an airplane's structure is three times its gross weight. Load factors are usually expressed in terms of "G"; that is, a load factor of 3 may be spoken of as 3 G's, and load factor of 4 as 4 G's, etc.

The stalling speed increases in proportion to the square root of the load factor. To further explain, the load factor produced in a 75° banked turn is 4. Applying the rule, the square root of 4 is 2. This means that an airplane with a normal unaccelerated stalling speed of 50 knots can be stalled at twice that speed or 100 knots, by inducing a load factor of 4.

HELICOPTER

1. When hovering, a helicopter tends to move in the direction of tail rotor thrust. This statement is

A—true; the movement is called transverse tendency.

B—true; the movement is called translating tendency.

C—false; the movement is opposite the direction of tail rotor thrust, and is called translating tendency.

Answer B—Subject Matter Knowledge Code: H71. AC 61-13B, Basic Helicopter Handbook; Chapter 2: Aerodynamics of Flight, page 15.

Translating tendency or drift. The entire helicopter has a tendency to move in the direction of tail rotor thrust (to the right) when hovering.

2. The purpose of lead-lag (drag) hinges in a three-bladed, fully articulated helicopter rotor system is to compensate for

Ab—Coriolis effect.

B—dissymmetry of lift.

C—blade flapping tendency.

Answer A—Subject Matter Knowledge Code: H71. AC 61-13B, Basic Helicopter Handbook; Chapter 5: Other Helicopter Components And Their Functions, page 32.

The purpose of the drag hinge and dampers is to absorb the acceleration and deceleration of the rotor blades caused by Coriolis effect.

GYROPLANE

1. To act as pilot in command of a gyroplane carrying passengers, what must the pilot accomplish in that gyroplane to meet recent daytime flight experience requirements?

A—Make nine takeoffs and nine landings within the preceding 30 days.

B—Make three takeoffs and three landings to a full stop within the preceding 90 days.

C—Make three takeoffs and three landings within the preceding 90 days.

Answer C—Subject Matter Knowledge Code: A20. 14 CFR part 61, section 61.57 states:

"Sec. 61.57 Recent flight experience: Pilot in command.

(a) General experience. (1) Except as provided in paragraph (e) of this section, no person may act as pilot in command of an aircraft carrying passengers or of an aircraft certificated for more than one pilot flight crewmember unless that person has made at least three takeoffs and three landings within the preceding 90 days, and—

(i) The person acted as the sole manipulator of the flight controls; and

(ii) The required takeoffs and landings were performed in an aircraft of the same category, class, and type (if a type rating is required), and, if the aircraft to be flown is an airplane with a tailwheel, the takeoffs and landings must have been made to a full stop in an airplane with a tailwheel.

(2) For the purpose of meeting the requirements of paragraph (a) (1) of this section, a person may act as a pilot in command of an aircraft under day VFR or day IFR, provided no persons or property are carried on board the aircraft, other than those necessary for the conduct of the flight...."

2. For gyroplanes with constant-speed propellers, the first indication of carburetor icing is usually

A—a decrease in engine RPM.

B—a decrease in manifold pressure.

C—engine roughness followed by a decrease in engine RPM.

Answer B—Subject Matter Knowledge Code H307. AC 61-23C, Pilot's Handbook of Aeronautical Knowledge; Chapter 2: Airplanes and Engines, page 2-11.

Indications of Carburetor Icing. For airplanes with fixed-pitch propellers, the first indication of carburetor icing is loss of RPM. For airplanes with controllable-pitch (constant-speed) propellers, the first indication is usually a drop in manifold pressure. In both cases, a roughness in engine operation may develop later. There will be no reduction in RPM in airplanes with constant-speed propellers, since propeller pitch is automatically adjusted to compensate for the loss of power, thus maintaining constant RPM.

GLIDER

1. What is the minimum age requirement for a person to be issued a student pilot certificate for the operation of gliders?

A—14 years of age.

B—15 years of age.

C—16 years of age.

Answer A—Subject Matter Knowledge Code: A22. 14 CFR part 61, section 61.83 states:

"Sec. 61.83 Eligibility requirements for student pilots.

To be eligible for a student pilot certificate, an applicant must:

(a) Be at least 16 years of age for other than the operation of a glider or balloon.

(b) Be at least 14 years of age for the operation of a glider or balloon.

(c) Be able to read, speak, write, and understand the English language. If the applicant is unable to meet one of these requirements due to medical reasons, then the Administrator may place such operating limitations on that applicant's pilot certificate as are necessary for the safe operation of the aircraft."

2. During aerotow of a glider that weighs 940 pounds, which towrope tensile strength would require the use of safety links at each end of the rope?

A—752 pounds.

B—1,500 pounds.

C—2,000 pounds.

Answer C—Subject Matter Knowledge Code: B12. 14 CFR part 91, section 91.309 states:

"Sec. 91.309 Towing: Gliders....

(3) The towline used has breaking strength not less than 80 percent of the maximum certificated operating weight of the glider and not more than twice this operating weight. However, the towline used may have a breaking strength more than twice the maximum certificated operating weight of the glider if—

(i) A safety link is installed at the point of attachment of the towline to the glider with a breaking strength not less than 80 percent of the maximum certificated operating weight of the glider and not greater than twice this operating weight.

(ii) A safety link is installed at the point of attachment of the towline to the towing aircraft with a breaking strength greater, but not more than 25 percent greater, than that of the safety link at the towed glider end of the towline and not greater than twice the maximum certificated operating weight of the glider;...."

Based on the last sentence in paragraph (3), twice the breaking strength of 940 pounds is 1,880 pounds. Any rope with a tensile strength of more than 1,880 pounds requires safety links as stated in (i) and (ii) of this section.

BALLOON—HOT AIR

1. Propane is preferred over butane for fuel in hot air balloons because

A—it has a higher boiling point.

B—it has a lower boiling point.

C—butane is very explosive under pressure.

Answer B—Subject Matter Knowledge Code: O171. Balloon Digest—Balloon Federation of America; Chapter 7: Chemical and Physical Properties, page 36.

Propane has a sustained boiling point of -44°F, which give a driving force to push it through the burner if the tanks (and their contained propane) are kept at any temperature above -44°F.

Butane is occasionally used as a fuel in balloons when propane is unavailable. It has a boiling temperature of only 32°F and therefore, at the same temperature, it doesn't have any where near as high a pressure development, decreasing the amount of fuel that can be driven through the burner and decreasing the performance of the balloon.

2. It may be possible to make changes in the direction of flight in a hot air balloon by

A—using the maneuvering vent.

B—operating at different flight altitudes.

C—flying a constant atmospheric pressure gradient.

Answer B—Subject Matter Knowledge Code: O263. How to Fly a Balloon—Balloon Publishing Company; page 13-7.

Balloon direction change usually comes with altitude change. Balloon pilots ascend and descend looking for different winds.

BALLOON—GAS

1. To perform a normal descent in a gas balloon, it is necessary to release

A—air.

B—gas.

C—ballast.

Answer B—Subject Matter Knowledge Code: O150. Balloon Digest; Chapter 2: Theory and Practice of Balloon Flight, page 9.

The gas balloon pilot can initiate a descent by releasing gas through a valve at the top of the envelope, and can then recover from the descent or initiate an ascent by releasing ballast.

2. What would cause a gas balloon to start a descent if a cold air mass is encountered and the envelope becomes cooled?

A—The expansion of the gas.

B—The contraction of the gas.

C—A barometric pressure differential.

Answer B—Subject Matter Knowledge Code: O150. Balloon Digest; Chapter 2: Theory and Practice of Balloon Flight, page 9.

As a gas balloon rises, it reaches an equilibrium altitude. It maintains this altitude until some change occurs. The most common change is in temperature, either because of the natural cooling encountered with altitude, a loss or gain of solar heating, or an ascent or descent initiated by the pilot.

LIGHTER-THAN-AIR—AIRSHIP

1. The pressure height with any airship is that height at which

A—both ballonets are empty.

B—both ballonets are inflated.

C—gas pressure is 3 inches of water.

Answer A—Subject Matter Knowledge Code: P01. Goodyear Airship Operations Manual; Chapter 4, page 4-42.

The "pressure height" of a nonrigid or semirigid airship is the height at which the ballonets become completely emptied of air and the envelope completely filled with gas, which is also determined by the percentage of gas fullness at the surface and atmospheric conditions.

2. For airship IFR operations off established airways, the ROUTE OF FLIGHT portion of an IFR flight plan should list VOR navigational aids which are no more than

A—40 miles apart.

B—70 miles apart.

C—80 miles apart.

Answer C—Subject Matter Knowledge Code: J15. AIM; Chapter 5: Air Traffic Control Procedures; Section 1: Preflight, paragraph 5-1-7.

FLIGHT PLAN -- IFR FLIGHTS — DIRECT FLIGHTS

The azimuth feature of VOR aids and the azimuth and distance (DME) features of VORTAC and TACAN aids are assigned certain frequency protected areas of airspace which are intended for application to established airway and route use, and to provide guidance for planning flights outside of established airways or routes. These areas of airspace are expressed in terms of cylindrical service volumes of specified dimensions called "class limits" or "categories." (Reference – NAVAID Service Volumes, paragraph 1-8). An operational service volume has been established for each class in which adequate signal coverage and frequency protection can be assured. To facilitate use of VOR, VORTAC, or TACAN aids, consistent with their operational service volume limits, pilot use of such aids for defining a direct route of flight in controlled airspace should not exceed the following:

(a) Operations above FL 450–Use aids not more than 200 NM apart. These aids are depicted in En Route High Altitude Charts.

(b) Operation off established routes from 18,000 feet MSL to FL 450–Use aids not more than 260 NM apart. These aids are depicted on En Route High Altitude Charts.

(c) Operation off established airways below 18,000 feet MSL–Use aids not more than 80 NM apart. These aids are depicted on En Route Low Altitude Charts.

(d) Operation off established airways between 14,500 feet MSL and 17,999 feet MSL in the conterminous U.S.–(H) facilities not more than 200 NM.

MILITARY COMPETENCE—AIRPLANE

1. If the operational category of an airplane is listed as utility, it would mean that this airplane could be operated in which of the following maneuvers?

A—Limited acrobatics, excluding spins.

B—Limited acrobatics, including spins (if approved).

C—Any maneuver except acrobatics or spins.

Answer B—Subject Matter Code: A150. 14 CFR part 23, section 23.3 states:

"Sec. 23.3 Airplane categories....

(3) Lazy eights, chandelles, and steep turns, in which the angle of bank is not more than 60 degrees.

(b) The utility category is limited to airplanes that have a seating configuration, excluding pilot seats, of nine or less, a maximum certificated takeoff weight of 12,500 pounds or less, and intended for limited acrobatic operation. Airplanes certificated in the utility category may be used in any of the operations covered under paragraph (a) of this section and in limited acrobatic operations. Limited acrobatic operation includes:

(1) Spins (if approved for the particular type of airplane); and

(2) Lazy eights, chandelles, and steep turns, or similar maneuvers, in which the angle of bank is more than 60 degrees but not more than 90 degrees."

2. To act as pilot in command of an airplane that is equipped with a retractable landing gear, flaps, and controllable pitch propeller, a person is required to

A—hold a multiengine airplane class rating.

B—make at least six takeoffs and landings in such an airplane within the preceding 6 months. C—receive flight instruction in such an airplane and obtain a logbook endorsement of competency.

Answer C—Subject Matter Knowledge Code: A20. 14 CFR part 61, section 61.31 states:

"Sec. 61.31 Type rating requirements, additional training, and authorization requirements....

(e) Additional training required for operating complex airplanes. (1) Except as provided in paragraph (e)(2) of this section, no person may act as pilot in command of a complex airplane (an airplane that has a retractable landing gear, flaps, and a controllable pitch propeller; or, in the case of a seaplane, flaps and a controllable pitch propeller), unless the person has—

(i) Received and logged ground and flight training from an authorized instructor in a complex airplane, or in a flight simulator or flight training device that is representative of a complex airplane, and has been found proficient in the operation and systems of the airplane; and

(ii) Received a one-time endorsement in the pilot's logbook from an authorized instructor who certifies the person is proficient to operate a complex airplane.

(2) The training and endorsement required by paragraph (e)(1) of this section is not required if the person has logged flight time as pilot in command of a complex airplane, or in a flight simulator or flight training device that is representative of a complex airplane prior to August 4, 1997...."

MILITARY COMPETENCE—HELICOPTER

1. While in flight a helicopter and an airplane are converging at a 90 angle, and the helicopter is located to the right of the airplane. Which aircraft has the right-of-way, and why?

A—The helicopter, because it is to the right of the airplane.

B—The helicopter, because helicopters have the right-of-way over airplanes.

C—The airplane, because airplanes have the right-of-way over helicopters.

Answer A—Subject Matter Knowledge Code: B08. 14 CFR part 91, section 91.113 states:

"Sec. 91.113 Right-of-way rules: Except water operations....

(d) Converging. When aircraft of the same category are converging at approximately the same altitude (except head-on, or nearly so), the aircraft to the other's right has the right-of-way. If the aircraft are of different categories—

(1) A balloon has the right-of-way over any other category of aircraft;

(2) A glider has the right-of-way over an airship, airplane, or rotorcraft; and

(3) An airship has the right-of-way over an airplane or rotorcraft...."

Although fixed and rotor-wing aircraft are a different category by definition, for the purpose of section 91.113 interpretation, they are considered to have similar maneuverability in forward, converging flight; especially when compared to aircraft of the different categories in (d), (1)(2) and (3) above.

2. What transponder equipment is required for helicopter operations within Class B airspace? A transponder

A—with 4096 code and Mode C capability.

B—as required for helicopter operations when visibility is less than 1 mile.

C—with 4096 code capability is required except when operating at or below 1,000 feet AGL under the terms of a letter of agreement.

Answer A—Subject Matter Code: B11. 14 CFR part 91, section 91.215 states:

"Sec. 91.215 ATC transponder and altitude reporting equipment and use...

(b) All airspace. Unless otherwise authorized or directed by ATC, no person may operate an aircraft in the airspace described in paragraphs (b)(1)

through (b)(5) of this section, unless that aircraft is equipped with an operable coded radar beacon transponder having either Mode 3/A 4096 code capability, replying to Mode 3/A interrogations with the code specified by ATC, or a Mode S capability, replying to Mode 3/A interrogations with the code specified by ATC and intermode and Mode S interrogations in accordance with the applicable provisions specified in TSO C112, and that aircraft is equipped with automatic pressure altitude reporting equipment having a Mode C capability that automatically replies to Mode C interrogations by transmitting pressure altitude information in 100-foot increments. This requirement applies—

(1) All aircraft. In Class A, Class B, and Class C airspace areas;...."

APPENDIX 1

LIST OF REFERENCE MATERIALS AND SUBJECT MATTER KNOWLEDGE CODES

The publications listed in the following pages contain study material you need to be familiar with when preparing for the commercial pilot knowledge tests. All of these publications can be purchased through U.S. Government Bookstores, commercial aviation supply houses, or industry organizations. The latest revision of the listed references should be requested. Additional study material is also available through these sources that may be helpful in preparing for commercial pilot knowledge tests. All publications listed would be excellent for a pilot to have in a personal reference library.

The subject matter knowledge codes refer to the specific reference for the knowledge standard. When reviewing results of your knowledge test, you should compare the subject matter knowledge code(s) on your Airman Test Report to the ones found below. This will be helpful for both review and preparation for the practical test.

Title 14 of the Code of Federal Regulations (14 CFR) part 1—Definitions and Abbreviations

A01	General Definitions
A02	Abbreviations and Symbols

14 CFR part 61—Certification: Pilots, Flight Instructors, and Ground Instructors

A20	General
A21	Aircraft Ratings and Pilot Authorizations
A22	Student Pilots
A24	Commercial Pilots
A26	Flight Instructors

14 CFR part 71—Designation of Class A, Class B, Class C, Class D, and Class E Airspace Areas; Airways; Routes; and Reporting Points

A60	General—Class A Airspace
A61	Class B Airspace
A64	Class C Airspace

14 CFR part 91— General Operating and Flight Rules

B07	General
B08	Flight Rules—General
B09	Visual Flight Rules
B10	Instrument Flight Rules
B11	Equipment, Instruments, and Certificate Requirements
B12	Special Flight Operations
B13	Maintenance, Preventive Maintenance, and Alterations

14 CFR part 125—Certification and Operations: Airplanes Having a Seating Capacity of 20 or More Passengers or a Maximum Payload Capacity of 6,000 Pounds or More

D30	General

NTSB 830—Rules Pertaining to the Notification and Reporting of Aircraft Accidents or Incidents and Overdue Aircraft, and Preservation of Aircraft Wreckage, Mail, Cargo, and Records

G10	General
G11	Initial Notification of Aircraft Accidents, Incidents, and Overdue Aircraft

AC 61-23—Pilot's Handbook of Aeronautical Knowledge

H300	Forces Acting on the Airplane in Flight
H301	Turning Tendency (Torque Effect)
H302	Airplane Stability
H303	Loads and Load Factors
H304	Airplane Structure
H305	Flight Control Systems
H306	Electrical System
H307	Engine Operation
H308	Propeller
H309	Starting the Engine
H310	Exhaust Gas Temperature Gauge
H311	Aircraft Documents, Maintenance, and Inspections
H312	The Pitot-Static System and Associated Instruments
H313	Gyroscopic Flight Instruments

H314	Magnetic Compass
H315	Weight Control
H316	Balance, Stability, and Center of Gravity
H317	Airplane Performance
H318	Observations
H319	Service Outlets
H320	Weather Briefings
H321	Nature of the Atmosphere
H322	The Cause of Atmospheric Circulation Appendix 1
H323	Moisture and Temperature
H324	Air Masses and Fronts
H325	Aviation Weather Reports, Forecasts, and Weather Charts
H326	Types of Airports
H327	Sources for Airport Data
H328	Airport Markings and Signs
H329	Airport Lighting
H330	Wind Direction Indicators
H331	Radio Communications
H332	Air Traffic Services
H333	Wake Turbulence
H334	Collision Avoidance
H335	Controlled Airspace
H336	Uncontrolled Airspace
H337	Special Use Airspace
H338	Other Airspace Areas
H339	Aeronautical Charts
H340	Latitude and Longitude
H341	Effect of Wind
H342	Basic Calculations
H343	Pilotage
H344	Dead Reckoning
H345	Flight Planning
H346	Charting the Course

H347	Filing a VFR Flight Plan
H348	Radio Navigation
H349	Obtaining a Medical Certificate
H350	Health Factors Affecting Pilot Performance
H351	Environmental Factors which Affect Pilot Performance

AC 91-23—Pilot's Weight and Balance Handbook

H10	Weight and Balance Control
H11	Terms and Definitions
H12	Empty Weight Center of Gravity
H13	Index and Graphic Limits
H14	Change of Weight
H15	Control of Loading—General Aviation
H16	Control of Loading—Large Aircraft

AC 60-14—Aviation Instructor's Handbook

H20	The Learning Process
H21	Human Behavior
H22	Effective Communication
H23	The Teaching Process
H24	Teaching Methods
H25	The Instructor as a Critic
H26	Evaluation
H27	Instructional Aids
H30	Flight Instructor Characteristics and Responsibilities
H31	Techniques of Flight Instruction
H32	Planning Instructional Activity

AC 61-21—Flight Training Handbook

H50	Introduction to Flight Training
H51	Introduction to Airplanes and Engines
H52	Introduction to the Basics of Flight
H53	The Effect and Use of Controls
H54	Ground Operations

H55	Basic Flight Maneuvers
H56	Airport Traffic Patterns and Operations
H57	Takeoffs and Departure Climbs
H58	Landing Approaches and Landings
H59	Faulty Approaches and Landings
H60	Proficiency Flight Maneuvers
H61	Cross-Country Flying
H62	Emergency Flight by Reference to Instruments
H63	Night Flying
H64	Seaplane Operations
H65	Transition to Other Airplanes
H66	Principles of Flight and Performance Characteristics

AC 61-13—Basic Helicopter Handbook

H70	General Aerodynamics
H71	Aerodynamics of Flight
H72	Loads and Load Factors
H73	Function of the Controls
H74	Other Helicopter Components and Their Functions
H75	Introduction to the Helicopter Flight Manual
H76	Weight and Balance
H77	Helicopter Performance
H78	Some Hazards of Helicopter Flight
H79	Precautionary Measures and Critical Conditions
H80	Helicopter Flight Maneuvers
H81	Confined Area, Pinnacle, and Ridgeline Operations
H82	Glossary

AC 61-27—Instrument Flying Handbook

I04	Basic Flight Instruments
I05	Attitude Instrument Flying—Airplanes
I08	Using the Navigation Instruments
I09	Radio Communications Facilities and Equipment
I10	The Federal Airways System and Controlled Airspace

AC 00-6—Aviation Weather

I20	The Earth's Atmosphere
I21	Temperature
I22	Atmospheric Pressure and Altimetry
I23	Wind
I24	Moisture, Cloud Formation, and Precipitation
I25	Stable and Unstable Air
I26	Clouds
I27	Air Masses and Fronts
I28	Turbulence
I29	Icing
I30	Thunderstorms
I31	Common IFR Producers
I32	High Altitude Weather
I33	Arctic Weather
I34	Tropical Weather
I35	Soaring Weather
I36	Glossary of Weather Terms

AC 00-45—Aviation Weather Services

I40	The Aviation Weather Service Program
I41	Surface Aviation Weather Reports
I42	Pilot and Radar Reports and Satellite Pictures
I43	Aviation Weather Forecasts
I44	Surface Analysis Chart
I45	Weather Depiction Chart
I46	Radar Summary Chart
I47	Significant Weather Prognostics
I48	Winds and Temperatures Aloft
I49	Composite Moisture Stability Chart
I50	Severe Weather Outlook Chart
I51	Constant Pressure Charts
I52	Tropopause Data Chart
I53	Tables and Conversion Graphs

AIM—Aeronautical Information Manual

J01	Air Navigation Radio Aids
J02	Radar Services and Procedures
J03	Airport Lighting Aids
J04	Air Navigation and Obstruction Lighting
J05	Airport Marking Aids and Signs
J06	Airspace—General
J07	Class G Airspace
J08	Controlled Airspace
J09	Special Use Airspace
J10	Other Airspace Areas
J11	Service Available to Pilots
J12	Radio Communications Phraseology and Techniques
J13	Airport Operations
J14	ATC Clearance/Separations
J15	Preflight
J16	Departure Procedures
J17	En Route Procedures
J18	Arrival Procedures
J19	Pilot/Controller Roles and Responsibilities
J20	National Security and Interception Procedures
J21	Emergency Procedures—General
J22	Emergency Services Available to Pilots
J23	Distress and Urgency Procedures
J24	Two-Way Radio Communications Failure
J25	Meteorology
J26	Altimeter Setting Procedures
J27	Wakc Turbulence
J28	Bird Hazards, and Flight Over National Refuges, Parks, and Forests
J29	Potential Flight Hazards
J30	Safety, Accident, and Hazard Reports
J31	Fitness for Flight
J32	Type of Charts Available

J33	Pilot Controller Glossary

Other Documents

J34	Airport/Facility Directory
J35	En Route Low Altitude Chart
J36	En Route High Altitude Chart
J37	Sectional Chart
J40	Standard Instrument Departure (SID) Chart
J41	Standard Terminal Arrival (STAR) Chart
J42	Instrument Approach Procedures
J43	Helicopter Route Chart

AC 67-2—Medical Handbook for Pilots

J52	Hypoxia
J53	Hyperventilation
J55	The Ears
J56	Alcohol
J57	Drugs and Flying
J58	Carbon Monoxide
J59	Vision
J60	Night Flight
J61	Cockpit Lighting
J62	Disorientation (Vertigo)
J63	Motion Sickness
J64	Fatigue
J65	Noise
J66	Age
J67	Some Psychological Aspects of Flying
J68	The Flying Passenger

ADDITIONAL ADVISORY CIRCULARS

K01	AC 00-24, Thunderstorms
K02	AC 00-30, Rules of Thumb for Avoiding or Minimizing Encounters with Clear Air Turbulence
K03	AC 00-34, Aircraft Ground Handling and Servicing
K04	AC 00-54, Pilot Wind Shear Guide

K11	AC 20-34, Prevention of Retractable Landing Gear Failure
K12	AC 20-32, Carbon Monoxide (CO) Contamination in Aircraft—Detection and Prevention
K13	AC 20-43, Aircraft Fuel Control
K20	AC 20-103, Aircraft Engine Crankshaft Failure
K40	AC 25-4, Inertial Navigation System (INS)
L05	AC 60-22, Aeronautical Decision Making
L10	AC 61-67, Stall and Spin Awareness Training
L15	AC 61-107, Operations of Aircraft at Altitudes Above 25,000 Feet MSL and/or MACH numbers (Mmo) Greater Than .75
L34	AC 90-48, Pilots' Role in Collision Avoidance
L42	AC 90-87, Helicopter Dynamic Rollover
L50	AC 91-6, Water, Slush, and Snow on the Runway
L52	AC 91-13, Cold Weather Operation of Aircraft
L53	AC 91-14, Altimeter Setting Sources
L57	AC 91-43, Unreliable Airspeed Indications
L59	AC 91-46, Gyroscopic Instruments—Good Operating Practices
L61	AC 91-50, Importance of Transponder Operation and Altitude Reporting
L62	AC 91-51, Airplane Deice and Anti-Ice Systems
L70	AC 91-67, Aircraft Ground Deicing
L80	AC 103-4, Hazard Associated with Sublimation of Solid Carbon Dioxide (Dry Ice) Aboard aircraft
L90	AC 105-2, Sport Parachute Jumping
M01	AC 120-12, Private Carriage Versus Common Carriage of Persons or Property.
M02	AC 120-27, Aircraft Weight and Balance Control
M08	AC 120-58, Pilot Guide—Large Aircraft Ground Deicing
M13	AC 121-195-1, Operational Landing Distances for Wet Runways; Transport Category Airplanes
M51	AC 20-117, Hazards Following Ground Deicing and Ground Operations in Conditions Conducive to Aircraft Icing

M52	AC 00-2, Advisory Circular Checklist

Soaring Flight Manual—Jeppesen-Sanderson, Inc.

N20	Sailplane Aerodynamics
N21	Performance Considerations
N22	Flight Instruments
N23	Weather for Soaring
N24	Medical Factors
N25	Flight Publications and Airspace
N26	Aeronautical Charts and Navigation
N27	Computations for Soaring
N28	Personal Equipment
N29	Preflight and Ground Operations
N30	Aerotow Launch Procedures
N31	Ground Launch Procedures
N32	Basic Flight Maneuvers and Traffic
N33	Soaring Techniques
N34	Cross-Country Soaring

Powerline Excerpts—Balloon Federation of America

O30	Excerpts

Goodyear Airship Operations Manual

P01	Buoyancy
P02	Aerodynamics
P03	Free Ballooning
P04	Aerostatics
P05	Envelope
P06	Car
P07	Powerplant
P08	Airship Ground Handling
P11	Operating Instructions
P12	History
P13	Training

NOTE: AC 00-2, Advisory Circular Checklist, transmits the status of all FAA advisory circulars (AC's), as well as FAA internal publications and miscellaneous flight information, such as Aeronautical Information Manual, Airport/Facility Directory, knowledge test guides, practical test standards, and other material directly related to a certificate or rating. AC 00-2 is accessible through the Internet at http://www.faa.gov/abc/ac-chklst/actoc.htm, or you may obtain a free copy from:

U.S. Department of Transportation

Subsequent Distribution Office, SVC-121.23

Ardmore East Business Center

3341 Q 75 Ave.

Landover, MD 20785

COMPUTER TESTING DESIGNEES

The following is a list of the computer testing designees authorized to give FAA airman knowledge tests. This list should be helpful in case you choose to register for a test or simply want more information.

Computer Assisted Testing Service (CATS)

1849 Old Bayshore Highway

Burlingame, CA 94010

Applicant inquiry and test registration: 1-800-947-4228

From outside the U.S. (650) 259-8550

Sylvan Prometric

1000 Lancaster Street

Baltimore, MD 21202

Applicant inquiry and test registration: 1-800-274-1900, 1-800-967-1100, or 1-800-359-3278

From outside the U.S. registrants should contact the appropriate Regional Service Center (RSC):

London, England RSC	44-181-607-9090
Paris, France RSC	33-1-4289-3122
Dusseldorf, Germany RSC	49-2159-9233-50
Tokyo, Japan RSC	813-3269-9620
Latin America RSC	(612) 820-5200

LaserGrade Computer Testing

16209 S.E. McGillivray, Suite L

Vancouver, WA 98683

Applicant inquiry and test registration: 1-800-211-2753 or 1-800-211-2754
From outside the U.S. (360) 896-9111

Made in United States
Troutdale, OR
11/03/2026

56844457R00024